Mr. Brown took our class to see a sky show. It was one of the greatest shows we have ever seen. We learned a lot about outer space.

We had to follow a man down a long hallway. We were shown into a large room with a round roof. The man made the room grow dark. Then the roof lit up with stars.

We were shown the moon and the stars. It was like seeing the starry sky from our own backyard. We learned how to follow the pattern of stars in the sky.

What do you know about stars? There are yellow stars, white stars, and other colors too. But no one knows how many stars there are.

The sun is a huge star that glows with heat and light. It is one of the hottest stars.

The sun is smaller than many other stars. It just looks big to us when seen from Earth. We should have known that!

We were shown Earth and the other planets. Earth is not the biggest planet. Other planets are much larger.

Earth

We were sorry when the sky show ended. But we had a lot of fun. And now we know a lot more about the night sky.

Do you want to see your own sky show? Just go outside at night!